KiCad - PL Editor Reference Manual

A catalogue record for this book is available from the Hong Kong Public Libraries.

Published in Hong Kong by Samurai Media Limited.

Email: info@samuraimedia.org

ISBN 978-988-8381-87-6

Background Cover Image by https://www.flickr.com/people/webtreatsetc/

Contents

1 Introduction to Pl_Editor

Pl_Editor is a page layout editor tool to create custom title blocks, and frame references.

The title block, associated to frame references, and other graphic items (logos) is called here a page layout

Basic page layout items are:

- **Lines**
- **Rectangles**
- **Texts** (with format symbols, that will be replaced by the actual text, like the date, page number…) in Eeschema or Pcbnew.
- **Poly-polygons** (mainly to place logos and special graphic shapes)
- **Bitmaps**.

Warning
Bitmaps can be plotted only by few plotters (PDF and PS only) Therefore, for other plotters, only a bounding box will be plotted.

- Items can be repeated, and texts and poly_polygons can be rotated.

2 Pl_Editor files

2.1 Input file and default title block

Pl_Editor reads or writes page layout description files *.kicad_wks (KiCad worksheet).

An internal default page layout description to display the default KiCad title block is used until a file is read

2.2 Output file

The current page layout description can be written in a ***.kicad_wks** file, using the S-expression format, which is widely used in KiCad.

This file can be used to show the custom page layout in Eeschema and/or Pcbnew.

3 Theory of operations

3.1 Basic page layout items properties:

Basic page layout items are:

- **Lines**
- **Rectangles**
- **Texts** (with format symbols, with will be replaced by the actual text, like the date, page number···) in Eeschema or Pcbnew.
- **Poly-polygons** (mainly to place logos and special graphic shapes). These poly polygons are created by **Bitmap2compon** and cannot be built inside pl_editor, because it is not possible to create such shapes by hand.
- **Bitmaps** to place logos.

Warning
Bitmaps can be plotted only by few plotters: PDF and PS only.

Therefore:

- **Texts, poly-polygons** and **bitmaps** are defined by a position, and can be rotated.
- **Lines** (in fact segments) and **rectangles** are defined by two points: a start point and a end point. They cannot be rotated (this is useless for segments)

These basic items can be repeated.

Texts which are repeated accept also an increment value for labels (has meaning only if the text is one letter or one digit)

3.2 Coordinates definition

Each position, start point and end point of items is always relative to a page corner.

This feature ensure you can define a page layout which is not dependent on the paper size.

3.3 Reference corners and coordinates:

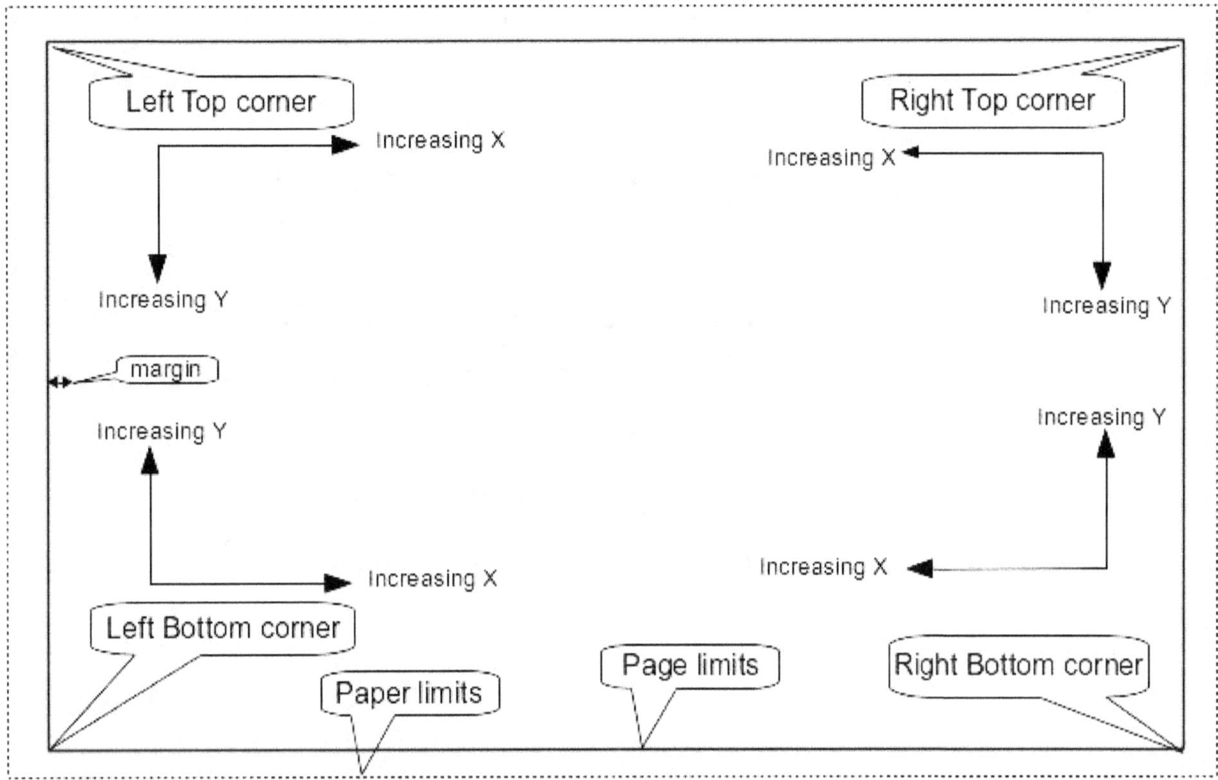

- When the page size is changed, the position of the item, relative to its reference corner does not change.
- Usually, title blocks are attached to the right bottom corner, and therefore this corner is the default corner, when creating an item.

For rectangles and segments, which have two defined points, each point has its reference corner.

3.4 Rotation

Items which have a position defined by just one point (texts and poly-polygons) can be rotated:

Normal: Rotation = 0

Rotated: Rotation = 20 and 10 degrees.

3.5 Repeat option

Items can be repeated:

This is useful to create grid and grid labels.

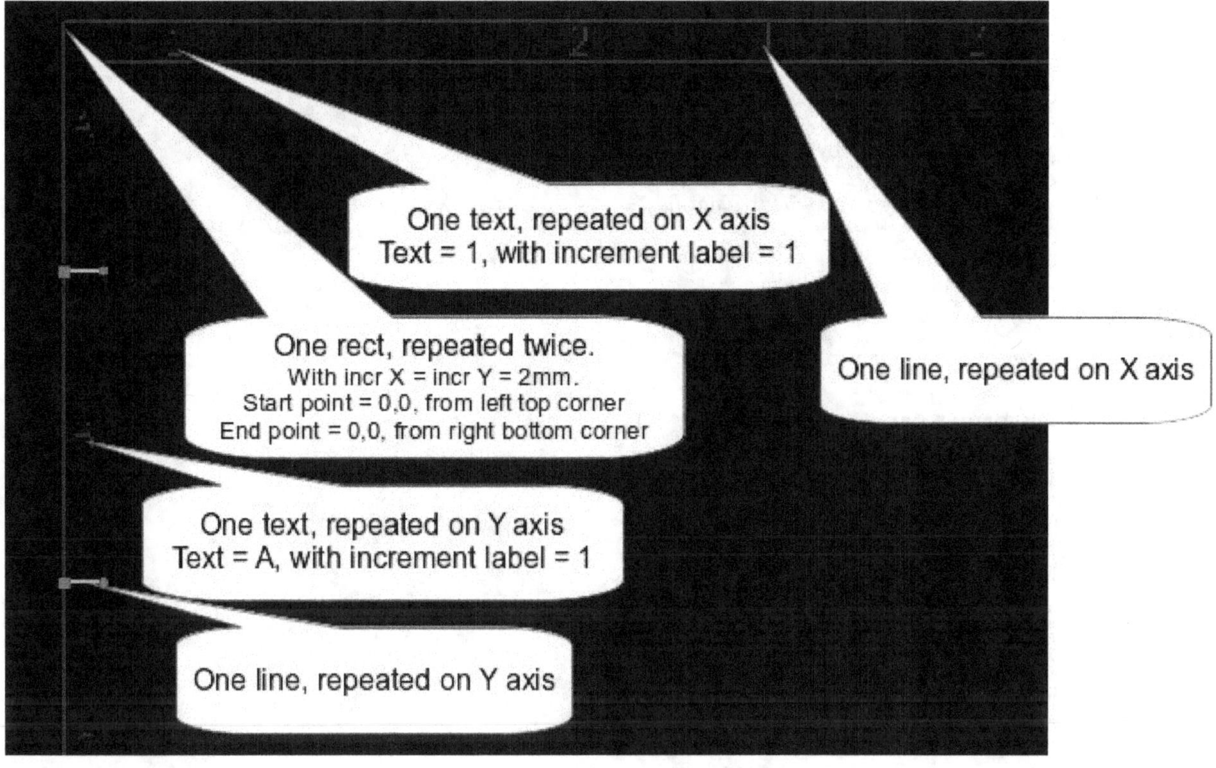

4 Texts and formats

4.1 Format symbols:

Texts can be simple strings or can include format symbols.

Format symbols are replaced by the actual values in Eeschema or Pcbnew.

They are like format symbols in printf function.

A format symbol is **%** followed by 1 letter.

The **%C** format has one digit (comment identifier)

Formats symbols are:

%% = replaced by %

%K = KiCad version

%Z = paper format name (A4, USLetter ···)

%Y = company name

%D = date

%R = revision

%S = sheet number

%N = number of sheets

%Cx = comment (x = 0 to 9 to identify the comment)

%F = filename

%P = sheet path (sheet full name, for Eeschema)

%T = title

Example:

"Size: %Z" displays "Size: A4" or "Size: USLetter"

User display mode: [T] activated. Title block displayed like in Eeschema and Pcbnew

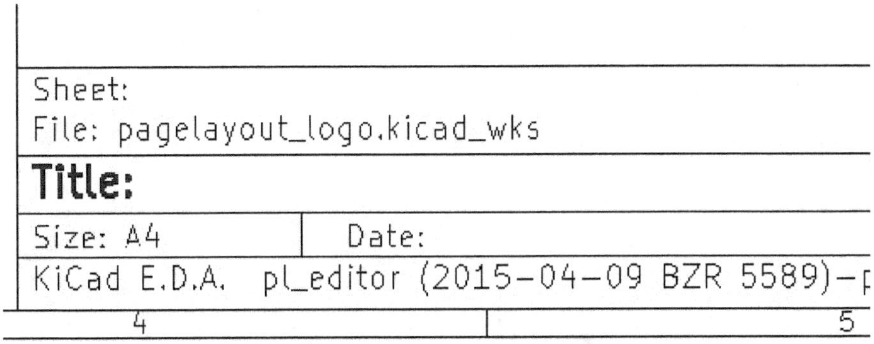

Sheet:
File: pagelayout_logo.kicad_wks
Title:
Size: A4 | Date:
KiCad E.D.A. pl_editor (2015-04-09 BZR 5589)-
4 | 5

"Native" display mode:  [%] activated. The native texts entered in Pl_Editor, with their format symbols.

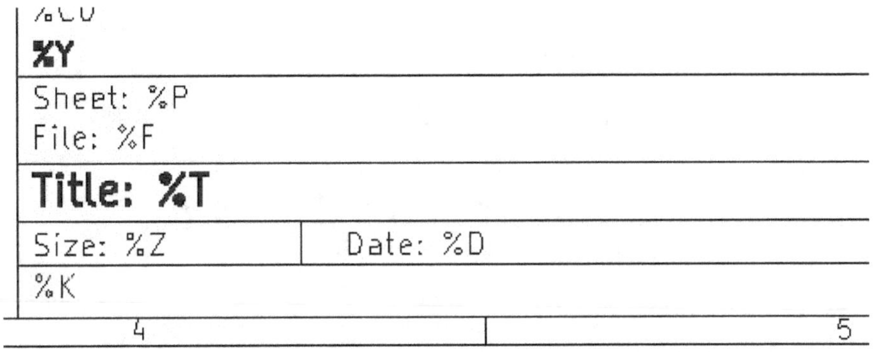

%Y
Sheet: %P
File: %F
Title: %T
Size: %Z | Date: %D
%K
4 | 5

4.2 Multi-line texts:

Texts can be multi-line.

There are 2 ways to insert a new line in texts:

1. Insert the "n" 2 chars sequence (mainly in Page setup dialog in KiCad)
2. Insert a new line in Pl_Editor Design window.

Here is an example:

Setup

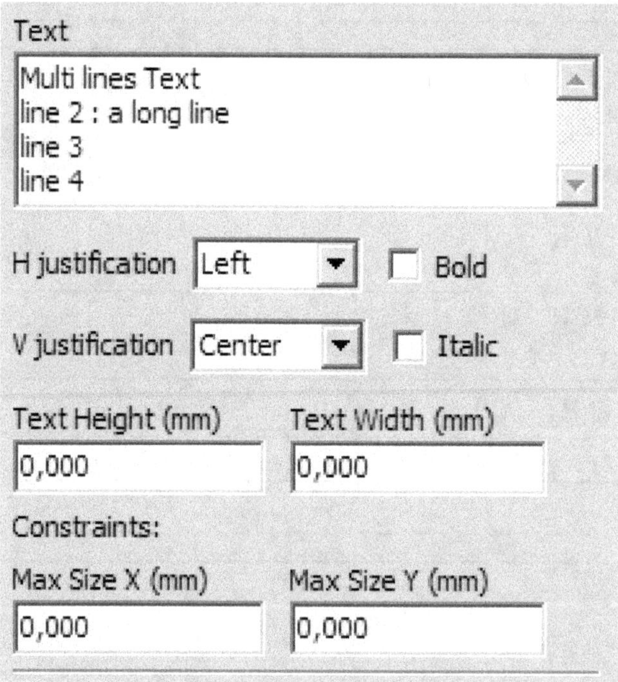

Output

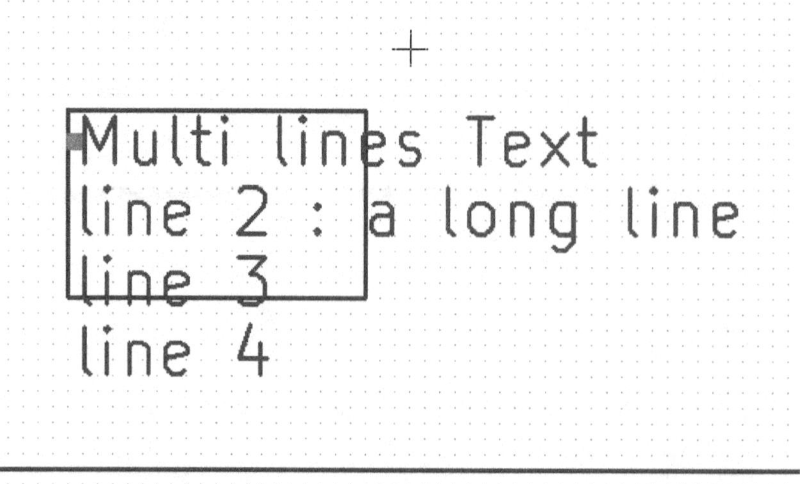

4.3 Multi-line texts in Page Setup dialog:

In the page setup dialog, text controls do not accept a multi-line text.

The "**\n**" 2 chars sequence should be inserted to force a new line inside a text.

Here is a two lines text, in *comment 2* field:

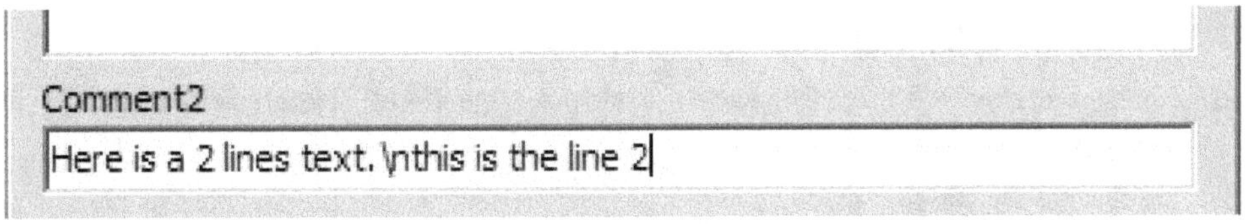

Comment2

Here is a 2 lines text. \nthis is the line 2

Here is the actual text:

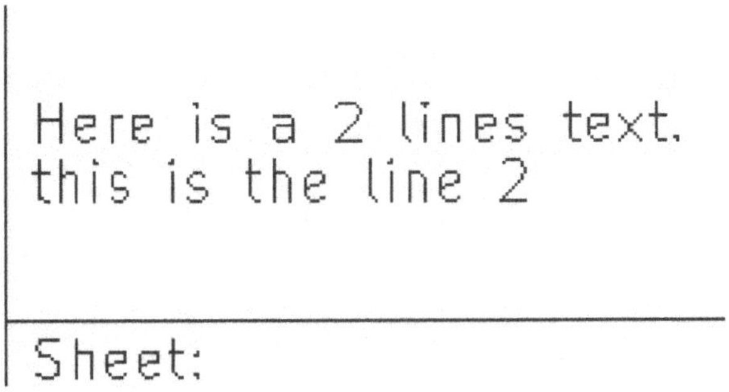

Here is a 2 lines text.
this is the line 2

Sheet:

However, if you really want the "**\n**" inside the text, enter "**\\n**".

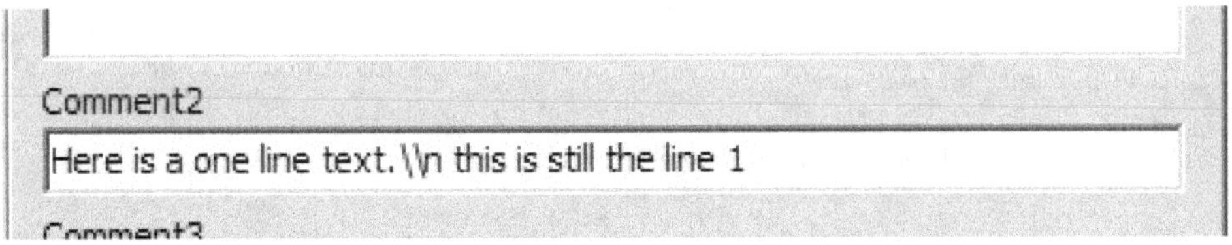

Comment2

Here is a one line text. \\n this is still the line 1

And the displayed text:

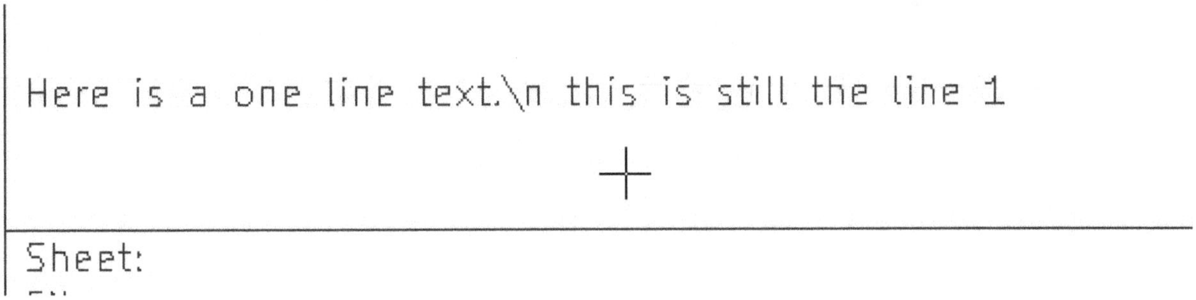

Here is a one line text.\n this is still the line 1

Sheet:

5 Constraints

5.1 Page 1 constraint

When using Eeschema, the full schematic often uses more than one page.

Usually page layout items are displayed on all pages.

But if a user want some items to be displayed only on page 1, or not on page 1, the "page 1 option" this is possible by setting this option:

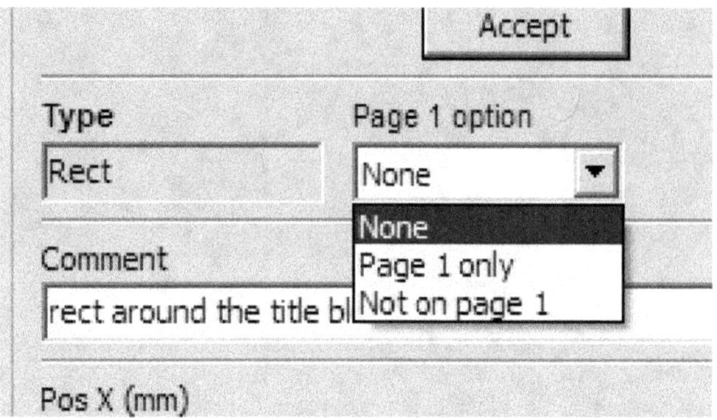

Page 1 option:

- None: no constraint.
- Page 1 only: the items is visible only on page 1.
- Not on page 1: the items is visible on all pages but the page 1.

5.2 Text full size constraint

Text

Multi lines Text
line 2 : a long line
line 3
line 4

H justification Left Bold

V justification Center Italic

Text Height (mm) 10,000 Text Width (mm) 0

Constraints:

Max Size X (mm) 10,000 Max Size Y (mm) 5

Only for texts, one can set 2 parameters :

- the max size X
- the max size Y

which define a bounding box

When these parameters are not 0, when displaying the text, the actual text height and the actual text width are dynamically modified if the full text size is bigger than the max size X and/or the max size Y, to fit the full text size with this bounding box.

When the actual full text size is smaller than the max size X and/or the max size Y, the text height and/or the text width is not modified.

The text with no bounding box. Max size X = 0,0 Max size Y = 0,0

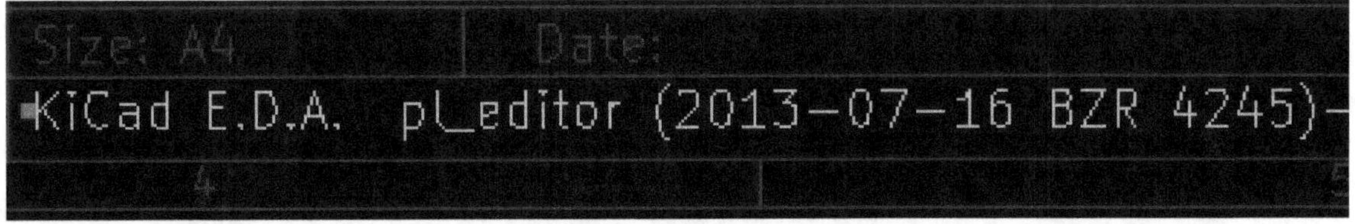

The **same** text with constraint. Max size X = 40,0 Max size Y = 0,0

A multi line text, constrained:

Setup

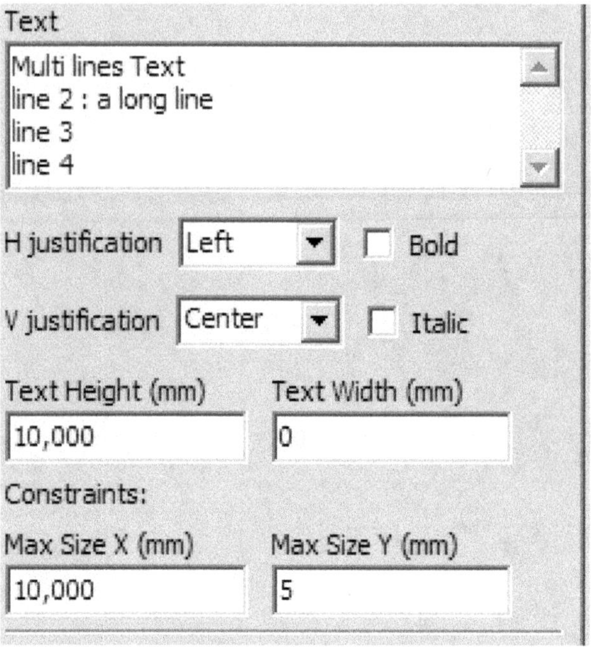

Output

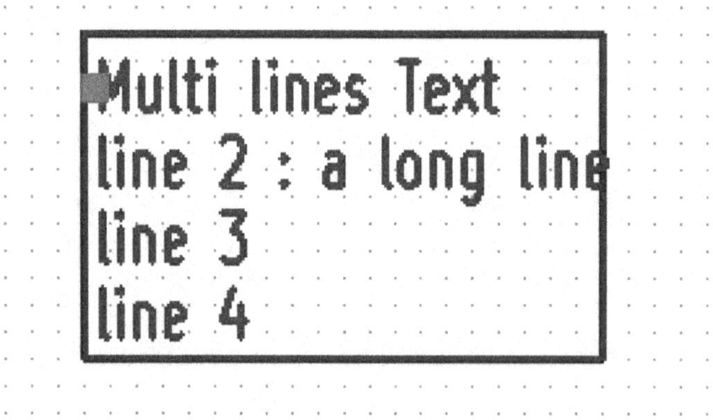

6 Invoking Pl_Editor

Pl_Editor is typically invoked from a command line, or from the KiCad manager.

From a command line, the syntax is pl_editor <*.kicad_wks file to open>.

7 Pl_Editor Commands

7.1 Main Screen

The image below shows the main window of Pl_Editor.

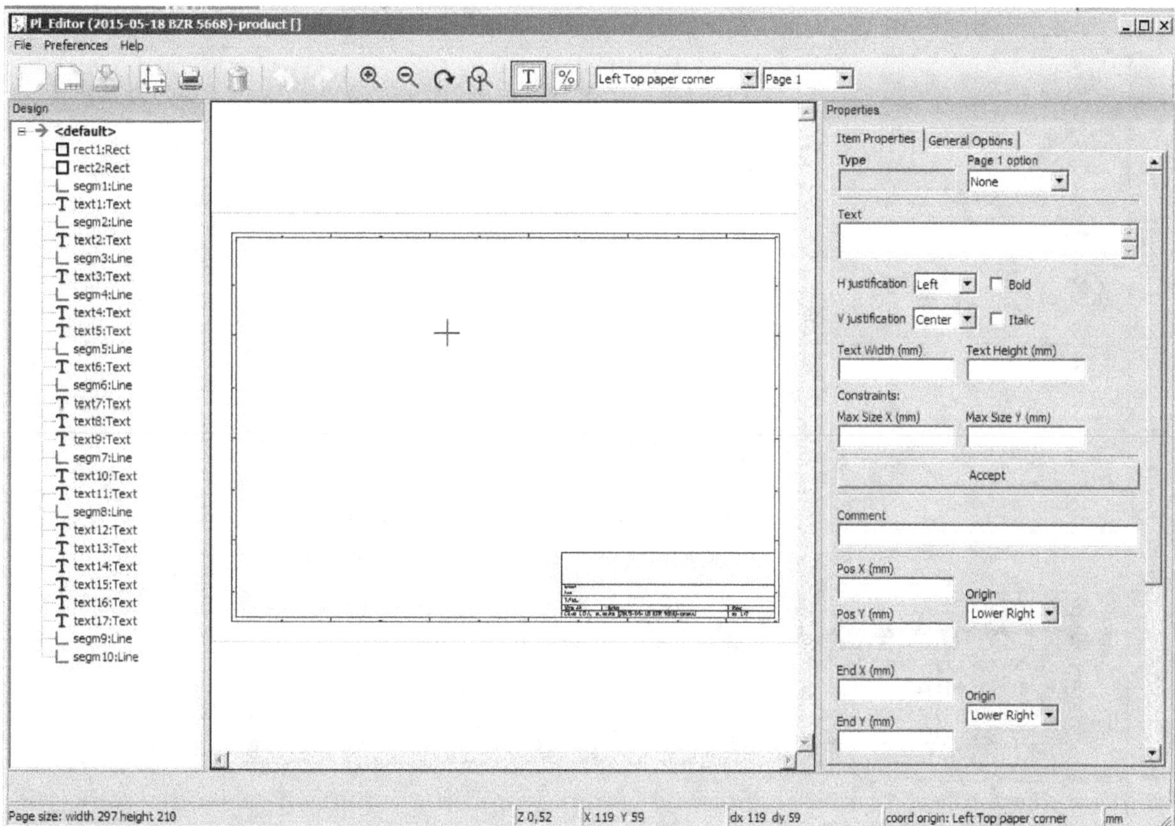

The left pane contains the list of basic items.

The right pane is the item settings editor.

7.2 Main Window Toolbar

The top toolbar allows for easy access to the following commands:

	Select the net list file to be processed.
	Load a page layout description file.
	Save the current page layout description in a .kicad_wks file.
	Display the page size selector and the title block user data editor.
	Prints the current page.
	Delete the currently selected item.
	Undo/redo tools.
	Zoom in, out, redraw and auto, respectively.
T	Show the page layout in user mode: texts are shown like in Eeschema or Pcbnew: text format symbols are replaced by the user texts.
%	Show the page layout in native mode: texts are displayed "as is", with the contained formats, without any replacement.
Left Top paper corner Left Top paper corner Right Bottom page corner Left Bottom page corner Right Top page corner Left Top page corner	Reference corner selection, for coordinates displayed to the status bar.
Page 1 Page 1 Other pages	Selection of the page number (page & or other pages). This selection has meaning only if some items than have a page option, are not shown on all pages (in a schematic for instance, which contains more than one page)

7.3 Commands in drawing area (draw panel)

7.3.1 Keyboard Commands

F1	Zoom In
F2	Zoom Out
F3	Refresh Display
F4	Move cursor to center of display window
Home	Fit footprint into display window
Space Bar	Set relative coordinates to the current cursor position
Right Arrow	Move cursor right one grid position
Left Arrow	Move cursor left one grid position
Up Arrow	Move cursor up one grid position
Down Arrow	Move cursor down one grid position

7.3.2 Mouse Commands

Scroll Wheel	Zoom in and out at the current cursor position
Ctrl + Scroll Wheel	Pan right and left
Shift + Scroll Wheel	Pan up and down
Right Button Click	Open context menu

7.3.3 Context Menu

Displayed by right-clicking the mouse:

- Add Line
- Add Rectangle
- Add Text
- Append Page Layout Descr File

Are commands to add a basic layout item to the current page layout description.

- Zoom selection: direct selection of the display zoom.
- Grid selection: direct selection of the grid.

Note

Append Page Layout Descr File is intended to add poly polygons to make logos.

Because usually a logo it needs hundred of vertices, you cannot create a polygon by hand. But you can append a description file, created by Bitmap2Component.

7.4 Status Bar Information

The status bar is located a the bottom of the Pl_Editor and provides useful information to the user.

Z 391,8 | X 40,9 Y 261 | dx -246 dy 60,9 | coord origin: Right Bottom page corner | mm

Coordinates are **always relative to the corner** selected as **reference**.

8 Left window

The left windows shows the list of layout items.

One can select a given item (left clicking on the line) or, when right clicking on the line, display a pop up menu.

This menu allows basic operations: add a new item, or delete the selected item.

→ **A selected item is also drawn in a different color on draw panel**.

Design tree: the item 19 is selected, and shown in highlighted on the draw panel.

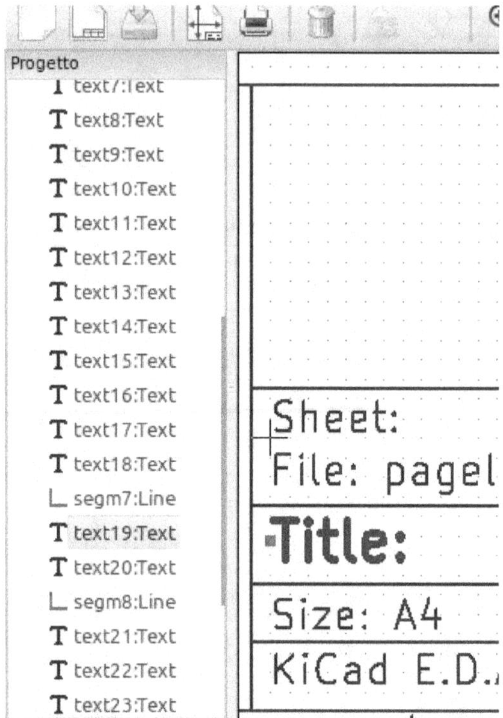

9 Right window

The right window is the edit window.

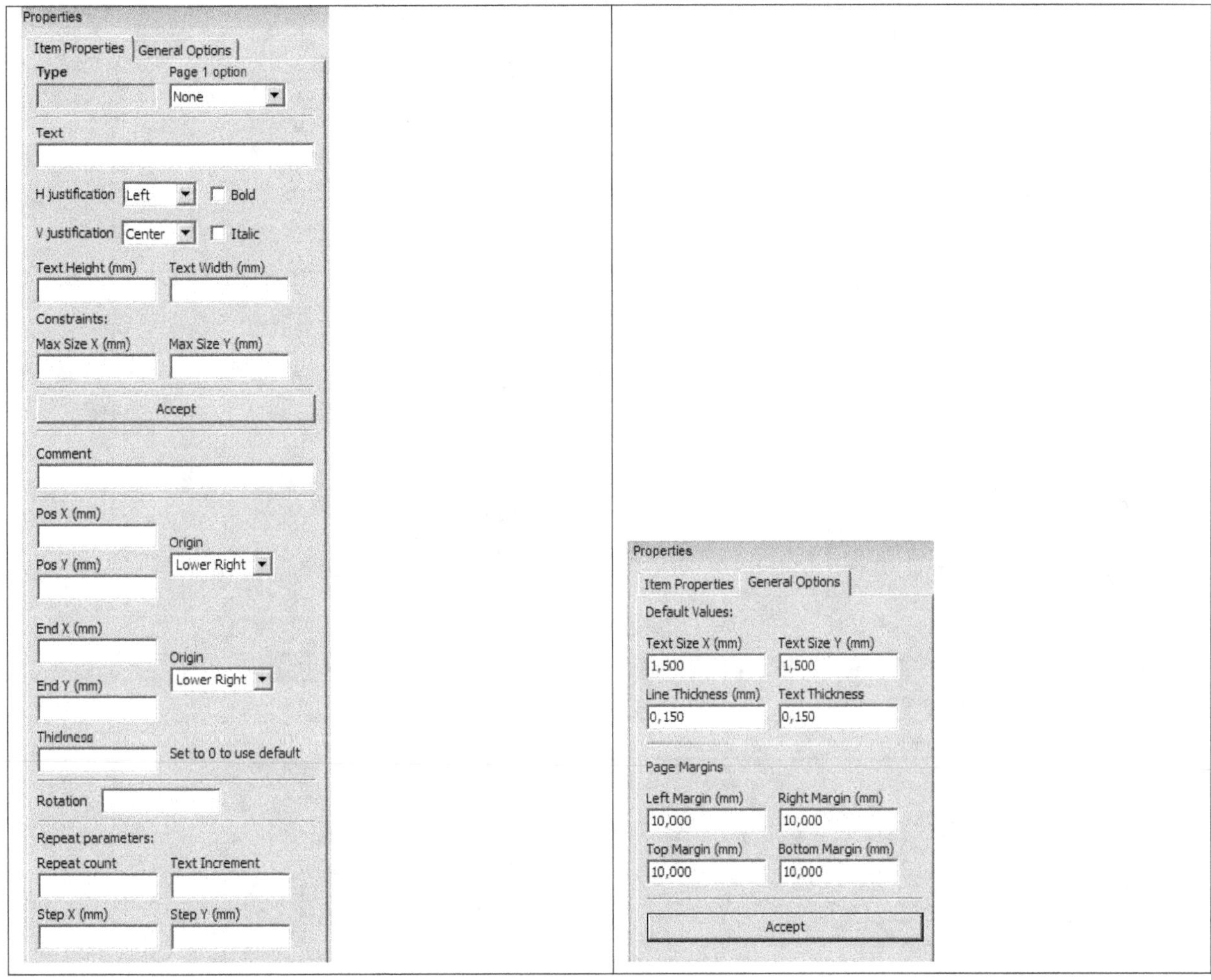

On this dialog you can set the page property and the item property of the current item.

Displayed settings depend on the selected item:

Settings for lines and rectangles	Settings for texts
Accept Type: Line — Page 1 option: None Comment Pos X (mm): 50,000 — Origin: Upper Left Pos Y (mm): 2,000 End X (mm): 50,000 — Origin: Upper Left End Y (mm): 0,000 Thickness: 0,000 — Set to 0 to use default Repeat parameters: Repeat count: 30 Step X (mm): 50,000 — Step Y (mm): 0,000	Accept Type: Text — Page 1 option: None Text: %K H justification: Left — Bold V justification: Center — Italic Text Height (mm): 0,000 — Text Width (mm): 0,000 Constraints: Max Size X (mm): 0,000 — Max Size Y (mm): 0,000 Comment Pos X (mm): 109,000 — Origin: Lower Right Pos Y (mm): 4,100 Thickness: 0,000 — Set to 0 to use default Rotation: 0,000 Repeat parameters: Repeat count: 1 — Text Increment: 1 Step X (mm): 0,000 — Step Y (mm): 0,000
Settings for poly-polygons	**Setting for bitmaps**
Accept Type: Poly — Page 1 option: Page 1 only Comment Pos X (mm): 136,002 — Origin: Lower Right Pos Y (mm): 18,002 Thickness: 0,010 Rotation: 20,000 Repeat parameters: Repeat count: 1 Step X (mm): 0,000 — Step Y (mm): 0,000	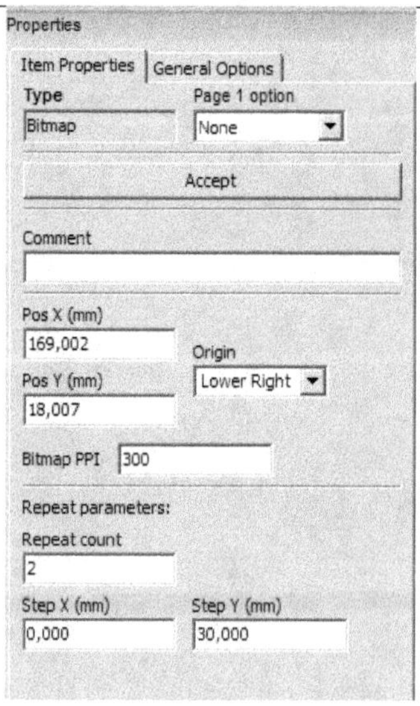 Properties Item Properties — General Options Type: Bitmap — Page 1 option: None Accept Comment Pos X (mm): 169,002 — Origin: Lower Right Pos Y (mm): 18,007 Bitmap PPI: 300 Repeat parameters: Repeat count: 2 Step X (mm): 0,000 — Step Y (mm): 30,000

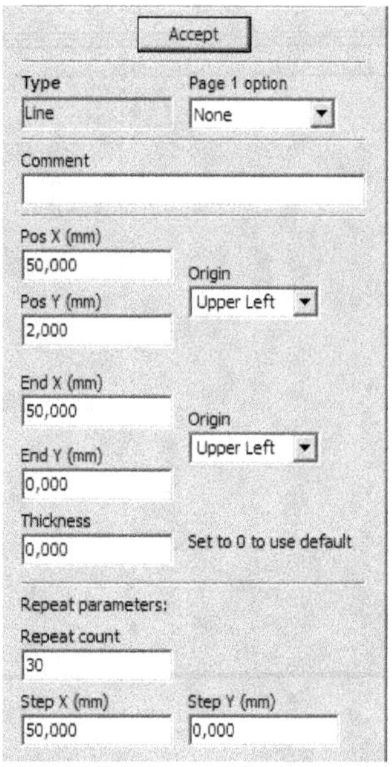

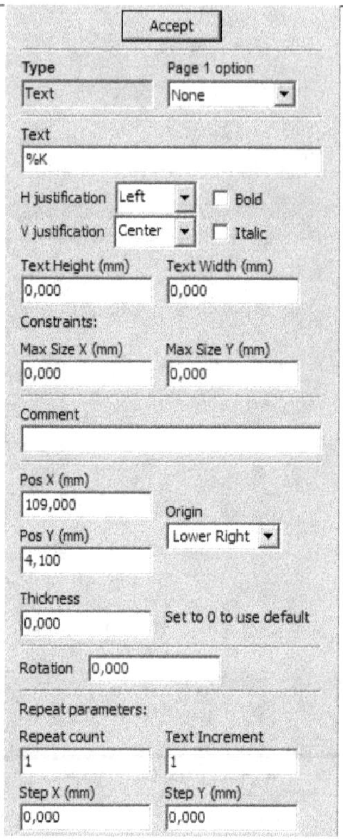

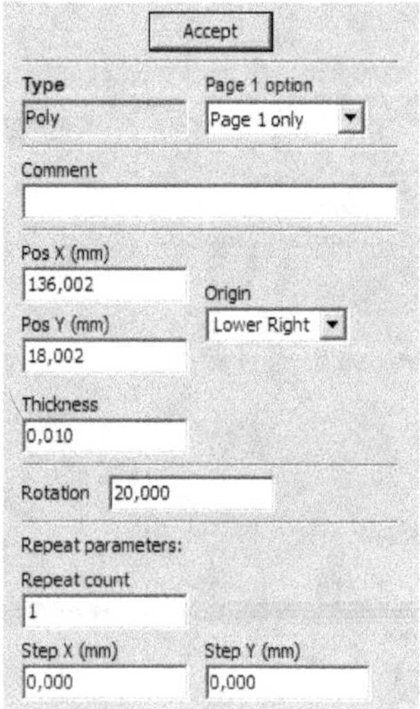

10 Interactive edition

10.1 Item selection

An item can be selected:

- From the Design tree.
- By Left clicking on it.
- By Right clicking on it (and a pop up menu will be displayed).

When selected, this item is drawn in yellow.

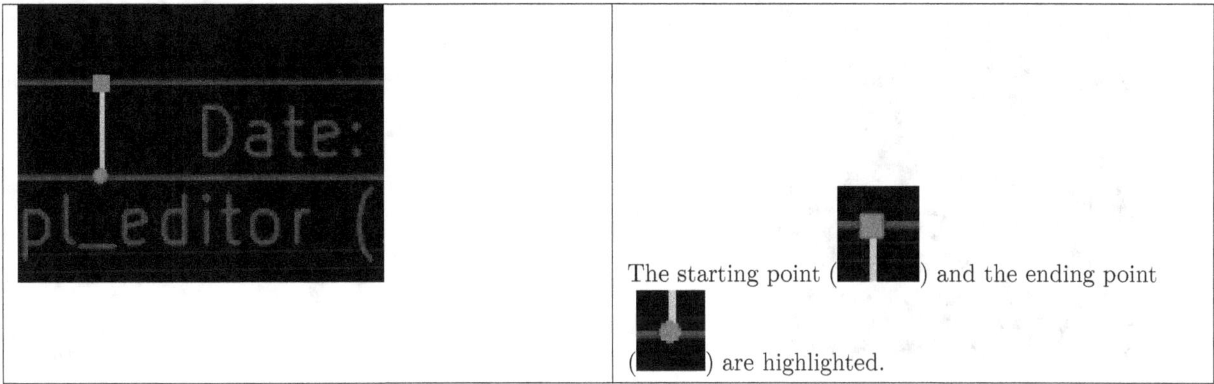

The starting point ( ) and the ending point ( ) are highlighted.

When right clicking on the item, a pop-up menu is displayed.

The pop menu options slightly depend on the selection:

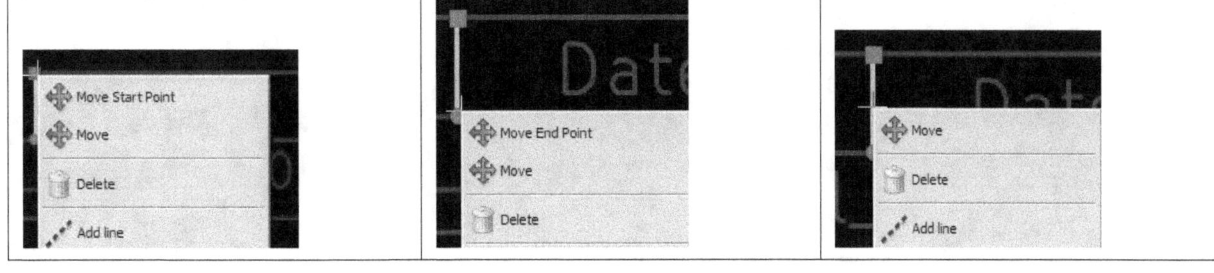

If more than one item is found, a menu clarification will be shown, to select the item:

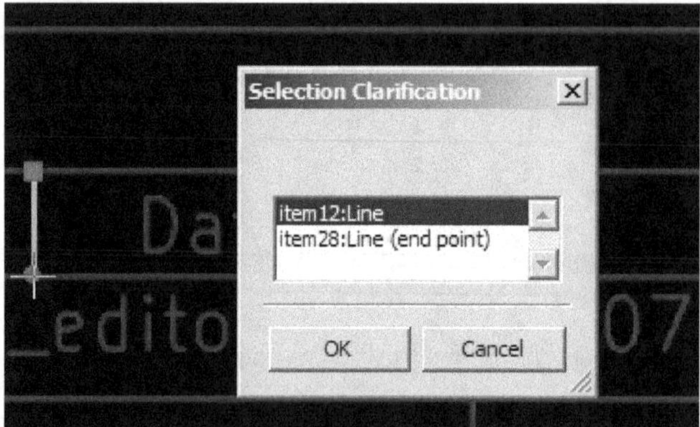

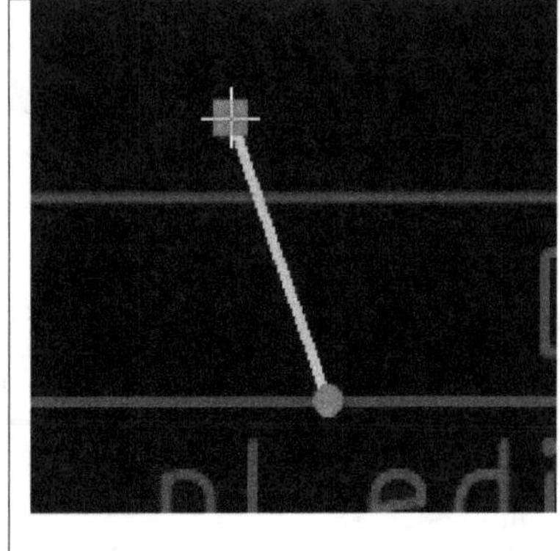	Once selected, the item, or one of its end points, can be moved by moving the mouse and placed (right clicking on the mouse).

10.2 Item creation

To add a new item, right click the mouse button when the cursor is on the left window or the draw area.

A popup menu is displayed:

Pop up menu in left window

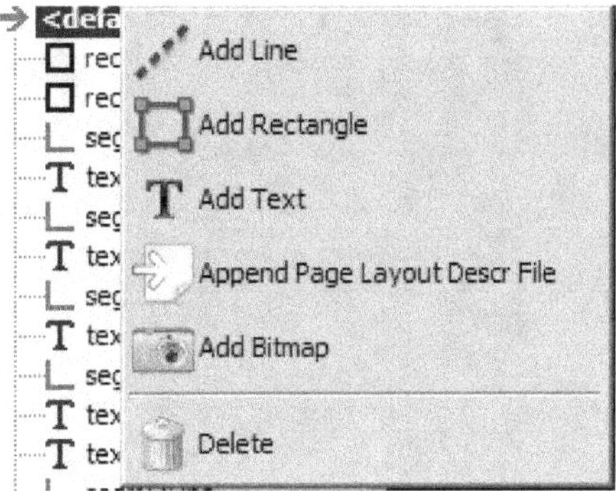

Pop up menu in draw area.

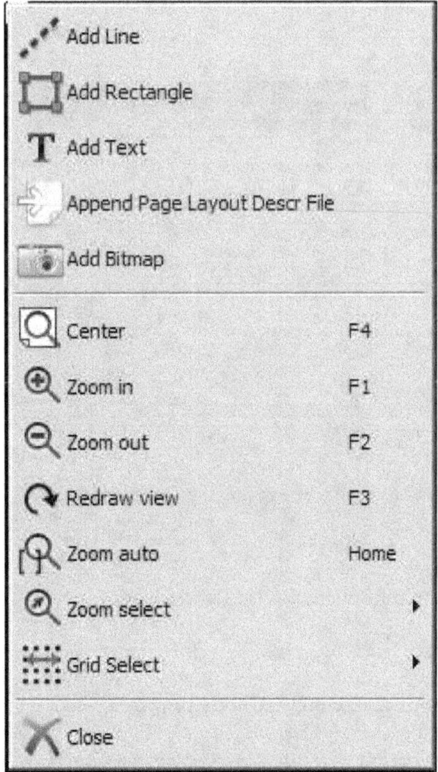

Lines, rectangles and texts are added just by clicking on the corresponding menu item.

Logos must first be created by Bitmap2component, which creates a page layout description file.

The Append Page Layout Descr File option append this file, to insert the logo (a poly polygon)

10.3 Adding lines, rectangles and texts

When clicking on the option, a dialog is opened:

Adding line or rectangle

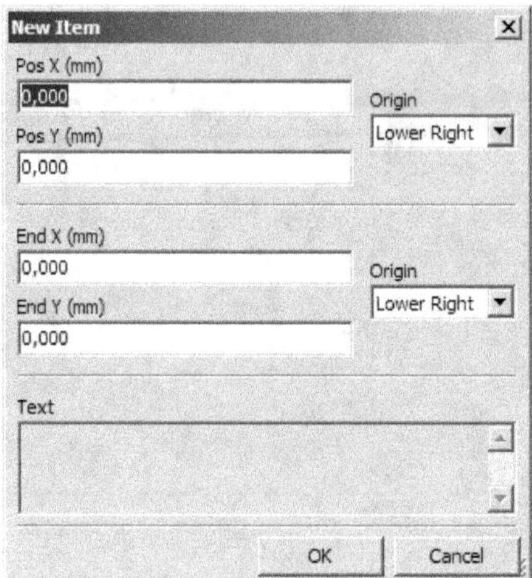

Adding text

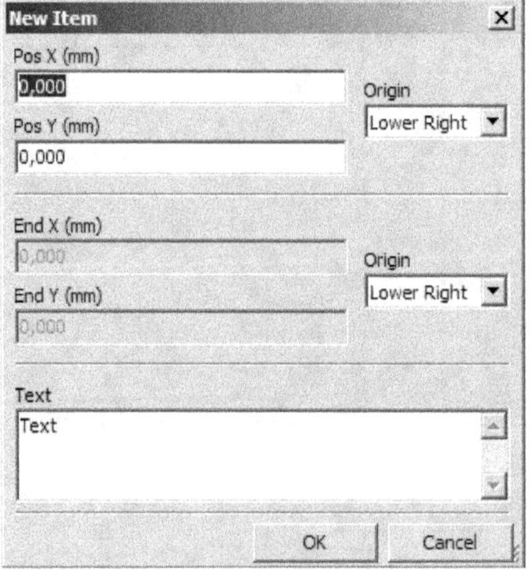

Position of end points, and corner reference can be defined here.

However they can be defined later, from the right window, or by moving the item, or one of its end points.

Most of time the corner reference is the same for both points.

If this is not the case, define the corner reference at creation is better, because if a corner reference is changed later, the geometry of the item will be a bit strange.

When an item is created, if is put in move mode, and you can refine its position (this is very useful for texts and small lines or rectangles)

10.4 Adding logos

To add a logo, a poly polygon (the vectored image of the logo) must be first created using Bitmap2component.

Bitmap2component creates a page layout description file which is append to the current design, using the **Append Page Layout Descr File** option.

Bitmap2component creates a page layout description file which contains only one item: a poly polygon.

However, this command can be used to append any page layout description file, which is merged with the current design.

Once a poly polygon is inserted, it can be moved and its parameters edited.

10.5 Adding image bitmaps

You can add an image bitmap using most of bitmap formats (PNG, JPEG, BMP ···)

- When a bitmap is imported, its PPI (pixel per inch) definition is set to 300PPI
- This value can be modified in panel Properties (right panel).
- The actual size depend on this parameter.
- Be aware that using higher definition values brings larger output files, and can have a noticeable draw or plot time.

A bitmap can be repeated, **but not rotated**.

www.ingramcontent.com/pod-product-compliance
Lightning Source LLC
Chambersburg PA
CBHW060009230526
45472CB00008B/2010

* 9 7 8 9 8 8 8 3 8 1 8 7 6 *